AF594053

THE WORLD'S SMARTEST ANIMALS
Chimpanzees
by Joanne Mattern
BLASTOFF! READERS 3
BELLWETHER MEDIA • MINNEAPOLIS, MN

Blastoff! Readers are carefully developed by literacy experts to build reading stamina and move students toward fluency by combining standards-based content with developmentally appropriate text.

Level 1 provides the most support through repetition of high-frequency words, light text, predictable sentence patterns, and strong visual support.

Level 2 offers early readers a bit more challenge through varied sentences, increased text load, and text-supportive special features.

Level 3 advances early-fluent readers toward fluency through increased text load, less reliance on photos, advancing concepts, longer sentences, and more complex special features.

★ Blastoff! Universe

Reading Level

Grade K

Grades 1–3

Grade 4

This edition first published in 2021 by Bellwether Media, Inc.

Library of Congress Cataloging-in-Publication Data

Names: Mattern, Joanne, 1963- author.
Title: Chimpanzees / by Joanne Mattern.
Description: Minneapolis, MN : Bellwether Media, 2021. | Series: Blastoff! readers: the world's smartest animals | Includes bibliographical references and index. | Audience: Ages 5-8 | Audience: Grades 2-3 | Summary: "Simple text and full-color photography introduce beginning readers to chimpanzees. Developed by literacy experts for students in kindergarten through third grade"-- Provided by publisher.
Identifiers: LCCN 2019059273 (print) | LCCN 2019059274 (ebook) | ISBN 9781644872376 (library binding) | ISBN 9781618919953 (ebook)
Subjects: LCSH: Chimpanzees--Behavior--Juvenile literature. | Animal intelligence--Juvenile literature.
Classification: LCC QL737.P94 M378 2021 (print) | LCC QL737.P94 (ebook) | DDC 599.88515/13--dc23
LC record available at https://lccn.loc.gov/2019059273
LC ebook record available at https://lccn.loc.gov/2019059274

Editor: Betsy Rathburn Designer: Jeffrey Kollock

Printed in the United States of America, North Mankato, MN.

Table of Contents

Intelligent Apes

Chimpanzees are **mammals** known for their **opposable thumbs**. They are often called chimps.

These animals are very **social**. They love to spend time with their friends!

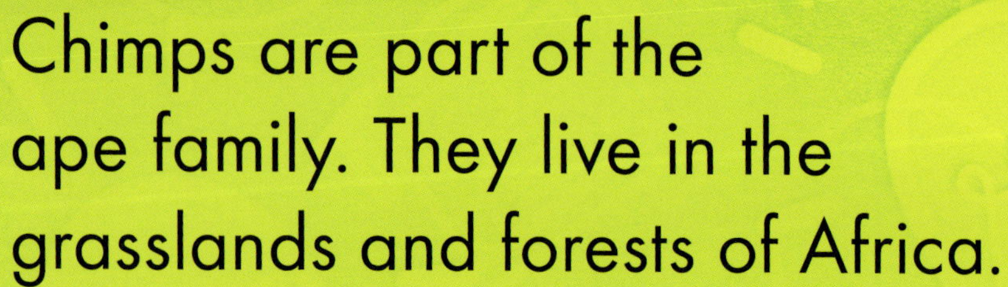
Chimps are part of the ape family. They live in the grasslands and forests of Africa.

Brain Size

human	chimpanzee
about 1,400 grams	about 420 grams

Chimps use **intelligence** to survive! They **communicate** with other chimps. They make tools to find food!

Talking and Tools

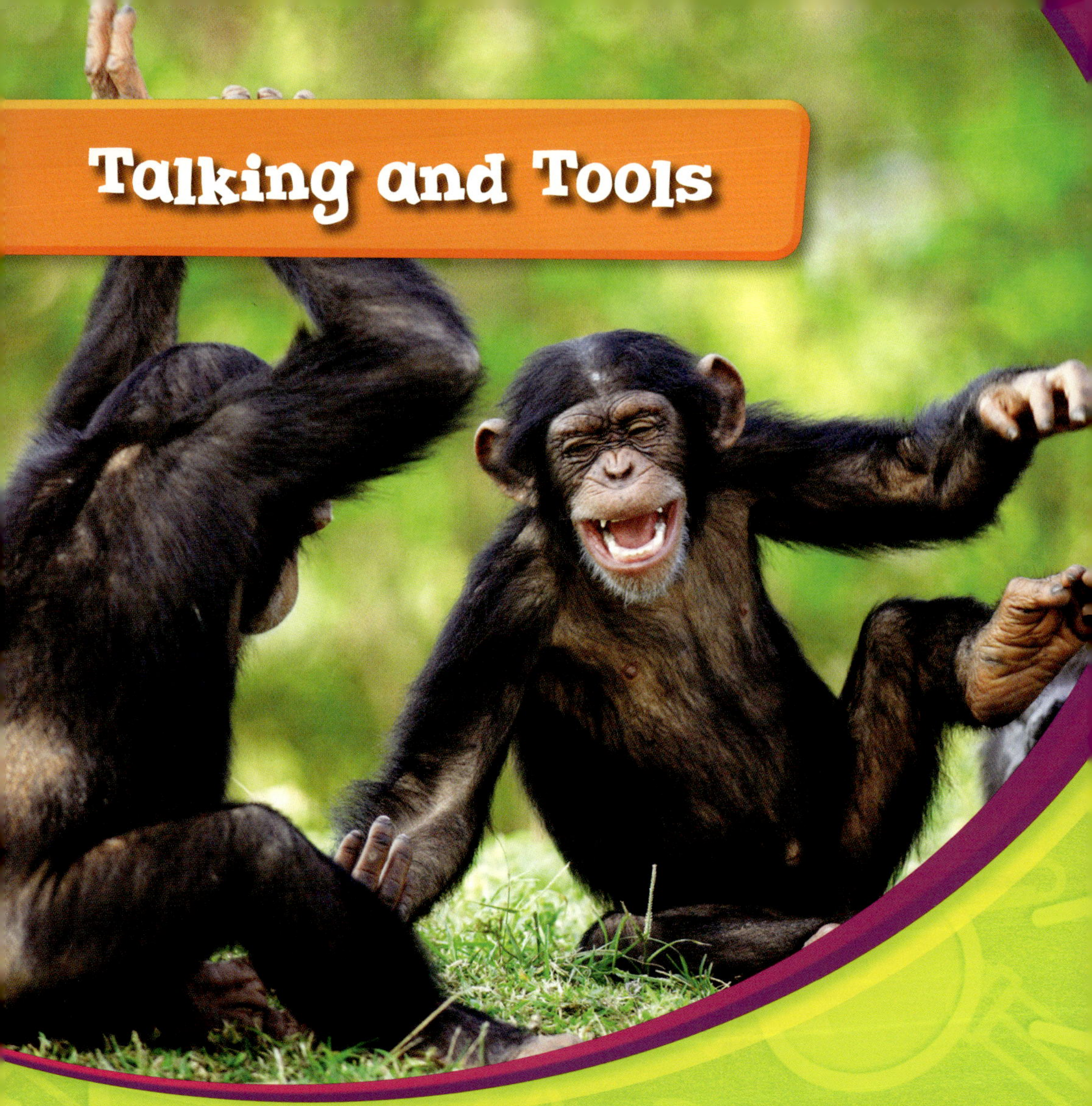

Chimps use their intelligence every day. They communicate using **gestures** and faces. They show their lower teeth when they want to play.

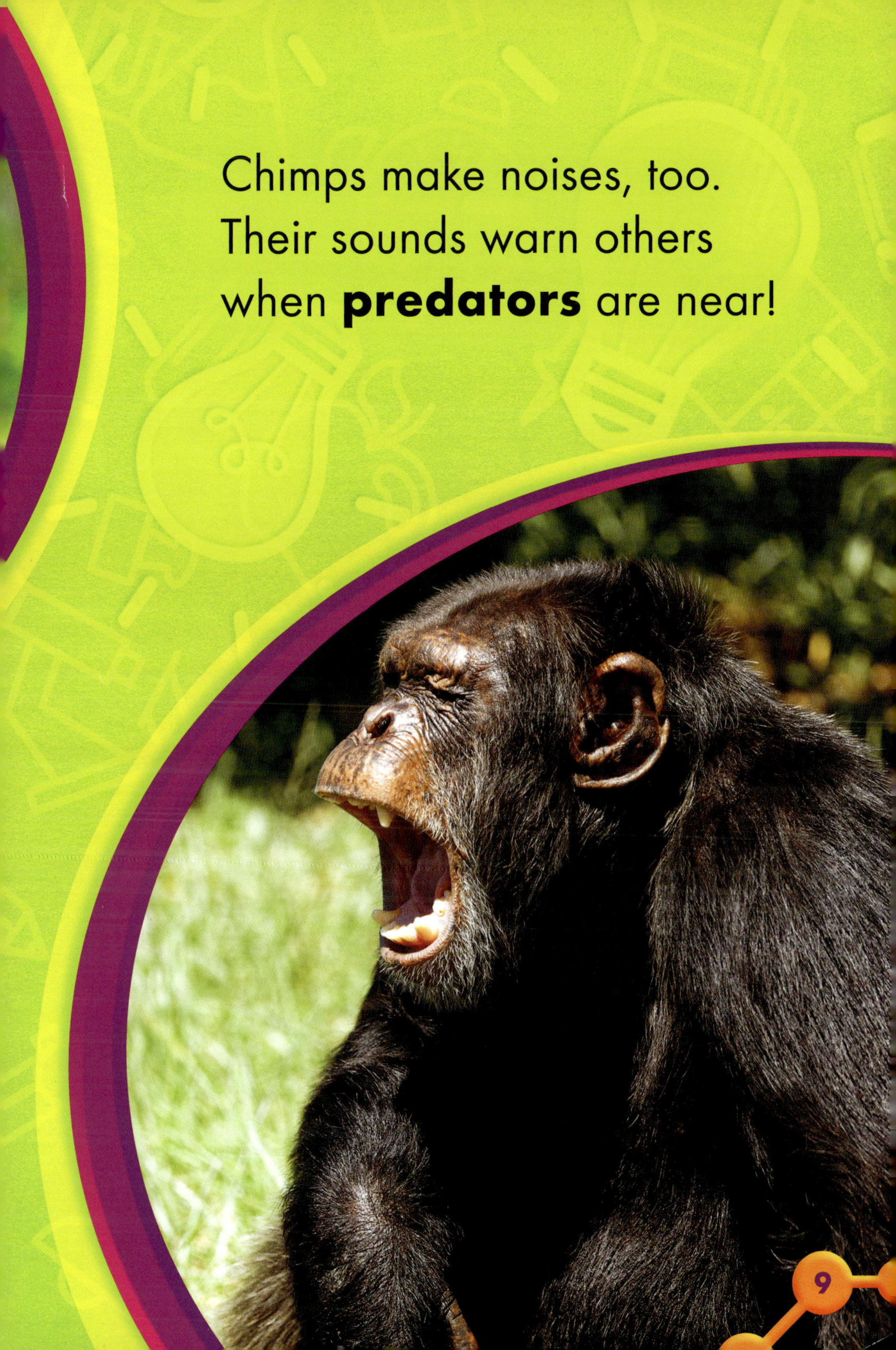

Chimps make noises, too. Their sounds warn others when **predators** are near!

Smart animals have **emotions**. Chimps do, too. They can feel sad or happy.

grooming

They even help other chimps when they are upset. They may hug or **groom** their friends.

These smart animals use tools. Sticks help chimps pull tasty **insects** from their nests. Sharp rocks break open nut shells.

Leaves are tools, too. They are held up as umbrellas. They are chewed as medicine!

Chimps have great memories, too. They remember chimps they have met in the past.

They remember places they have been before, too. This helps them find friends and food!

Learning About Chimpanzees

Many scientists have studied chimpanzees in the wild. Jane Goodall is the most famous.

She lived with chimps for years. Her **research** showed that chimps are intelligent!

Amazing Chimpanzee

Name

- Ham

Species

- common chimpanzee

Famous For

- First chimpanzee in space
- Learned to pull levers and take tests in the spacecraft
- Made it possible for humans to go to space

Chimpanzee Study

Question

Do chimpanzees have good memories?

Process

1. Chimpanzees were shown the numbers 1 to 9 on a touch screen.
2. Then, only the lowest number appeared on the screen.
3. The chimpanzees had to touch the spaces where the other numbers had been. They had to touch in order from lowest to highest.

What Happened?

- The chimpanzees pressed the numbers in the correct order.
- The chimpanzees could remember where the numbers were placed, even when they could not see the numbers.

Answer

- Chimpanzees have good memories.

Scientists research chimps in **labs**. Some studies show that chimps enjoy solving puzzles. Other studies show that chimps can learn. Many have learned **sign language**!

Studying chimps helps humans. It teaches scientists about human intelligence. They learn how human language started.

Chimpanzees are intelligent animals. They can teach people many things!

Glossary

communicate—to share thoughts and feelings using sounds, faces, and actions

emotions—feelings

gestures—actions that tell information

groom—to clean or make neat

insects—small animals with six legs and hard outer bodies; an insect's body is divided into three parts.

intelligence—the ability to learn or understand

labs—places where scientists research

mammals—warm-blooded animals that have backbones and feed their young milk

opposable thumbs—thumbs that can touch the other fingers and toes so that hands and feet can grasp and hold things

predators—animals that hunt other animals for food

research—careful study to learn new information

sign language—a language made up of hand gestures

social—marked by having many relationships with others

To Learn More

AT THE LIBRARY

Dickmann, Nancy. *Chimpanzees*. Tucson, Ariz.: Brown Bear Books, 2019.

Hansen, Grace. *Jane Goodall: Chimpanzee Expert & Activist*. Minneapolis, Minn.: Abdo Kids, 2015.

Kopp, Megan. *The Language of Chimpanzees and Other Primates*. New York, N.Y.: Cavendish Square, 2017.

ON THE WEB

FACTSURFER

Factsurfer.com gives you a safe, fun way to find more information.

1. Go to www.factsurfer.com.
2. Enter "chimpanzees" into the search box and click 🔍.
3. Select your book cover to see a list of related content.

Index

The images in this book are reproduced through the courtesy of: Eric Isselee, front cover, pp. 1, 14; Fly_and_Dive, p. 3; Abeselom Zerit, pp. 4, 5, 23; Eric Gevaert, p. 5; apple2499, pp. 6, 7; Arco Images GmbH/ Alamy, pp. 8, 9; Beate Wolter, p. 9; Sergey Uryadnikov, pp. 10, 11; Thomas Marent/ SuperStock, p. 11; lightofchairat, p. 12; Steve Bloom Images/ Alamy, pp. 12, 13; Andrew Lilley, pp. 14, 15; Suzi Eszterhas/ SuperStock, pp. 16, 17; Ferenc Szelepcsenyi, p. 16; Marshall Space Flight Center's Marshall Image Exchange/ Wikipedia, p. 17; Nature Picture Library/ Alamy, p. 19; CherylRamalho, pp. 20, 21.